Problem Solver II™

Student Workbook

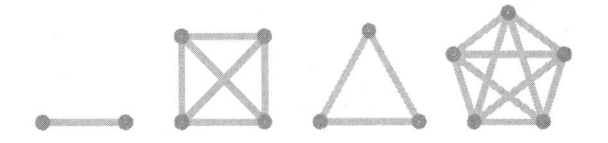

Judy Goodnow
Shirley Hoogeboom

McGraw Hill Wright Group

Acknowledgments

We wish to thank Marj Santos for reviewing the manuscript and guiding the classroom testing.

Judy Goodnow has authored and coauthored over 100 books and software programs for mathematics and problem solving. She has taught children from kindergarten through sixth grade. She holds a bachelor of arts degree from Wellesley College, a master's degree from Stanford University, and a California Teaching Credential from San Jose State University.

Shirley Hoogeboom has authored and coauthored over 100 books for mathematics and language arts. She has been a classroom teacher, and has conducted workshops for teachers on problem solving and on using math manipulatives. She holds a bachelor of arts degree in Education from Calvin College, where she earned a Michigan Teaching Credential. She completed further studies at California State University, Hayward, where she earned a California Teaching Credential.

McGraw Hill Wright Group

Problem Solver II: Integrating Problem Solving with Your Math Curriculum, Student Workbook, Grade 6
Copyright ©2005 Wright Group/McGraw-Hill
Text by Judy Goodnow and Shirley Hoogeboom
Illustrations by John Haslam
Design and Production by O'Connor Design

Problem Solver™ is a trademark of the McGraw-Hill Companies, Inc.

Published by Wright Group/McGraw-Hill, a division of the McGraw-Hill Companies, Inc. All rights reserved. No part of this publication may be reproduced or distributed in any form or by any means, or stored in a database or retrieval system, without the prior written consent of Wright Group/McGraw-Hill, including, but not limited to, network or other electronic storage or transmission, or broadcast for distance learning.

Wright Group/McGraw-Hill
One Prudential Plaza
Chicago, Illinois 60601
www.WrightGroup.com
Customer Service 888-772-4543

Printed in the United States of America.

1 2 3 4 5 6 7 8 9 MAL 10 09 08 07 06 05 04

ISBN: 0-322-08817-8

The McGraw-Hill Companies

Problem-Solving Strategies

Act Out or Use Objects

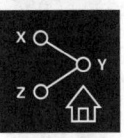

Use or Make a Picture or Diagram

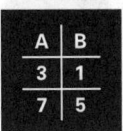

Use or Make a Table

Make an Organized List

Guess and Check

Use or Look for a Pattern

Work Backwards

Use Logical Reasoning

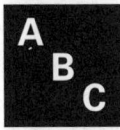

Make It Simpler

Brainstorm

Act Out or Use Objects

Play Money

1 Seth gave Marty this puzzle to solve:
Start with 10 dimes, 6 nickels, 5 $1 bills, and 5 $5 bills. Take away coins and bills until you have between $18 and $20 left, and:
- The value of the dimes is 20% of the value of the $1 bills.
- The value of the nickels is 25% of the value of the dimes.
- The value of the $1 bills is 20% of the value of the $5 bills.

When Marty solved this puzzle, what coins and bills did he have left?

FIND OUT
What do you have to find out to solve the problem?

What does the problem tell you about the coins and bills Marty has?

What does he have to do with the coins and bills?

CHOOSE STRATEGIES
You can **Act Out or Use Objects** to help you solve this kind of problem. Use play money to help you think about the problem and work out a solution.

Problem Solver II

SOLVE IT

1. What is the value of the coins and bills that Marty started with?

2. Begin with the $5 bills. How many do you want to take away?

   What is the value of the $5 bills now?

   Do the $1 bills show 20% of this new value?

   Can you take away $1 bills to show 20% of this new value?

   What would be the total value of the coins and bills now?

3. Do you want to take away another $5 bill?

   Do the $1 bills show 20% of the new amount?

   What do you need to take away from the $1 bills?

4. Now look at the dimes and $1 bills. Do the dimes show 20% of the value of the $1 bills?

   How can you change this?

5. Now compare the dimes and nickels. Is the value of the nickels 25% of the value of the dimes?

   How can you change that?

6. When Marty solved the puzzle, what coins and bills did he have left?

LOOK BACK
- Read the problem again.
- Check over your work.
- Did you answer the question that was asked?
- Does your answer make sense?

Act Out or Use Objects

Play Money

2 **Allie picked a card in the Money Challenge game that said: Start with 8 quarters, 10 $1 bills, 5 $5 bills, 5 $10 bills, and 2 $20 bills. Take away coins and bills until the total value of the money left is between $90 and $95, and:**
- **The value of the quarters is $\frac{1}{5}$ of the value of the $1 bills.**
- **The value of the $1 bills is $\frac{1}{3}$ of the value of the $5 bills.**
- **The value of the $5 bills is 50% of the value of the $10 bills.**
- **The value of the $10 bills is 75% of the value of the $20 bills.**

When Allie solved the challenge, what coins and bills did she have left?

FIND OUT

What do you have to find out to solve the problem?

What does the problem tell you about the coins and bills Allie started with?

What does she have to do with the coins and bills?

CHOOSE STRATEGIES

You can **Act Out or Use Objects** to help you solve this kind of problem. Use play money to help you think about the problem and work out a solution.

Problem Solver II

SOLVE IT

1. What is the value of the coins and bills that Allie started with?

   To solve the challenge, what should the total be?

2. Begin with the $20 bills. How many do you want to take away?

   What will the value of the $20 bills be now?

   Do the $10 bills show 75% of this new value?

   Can you take away $10 bills to show 75% of this new value?

3. Start over again. Look at the $10 bills and the $20 bills.

   What can you take away to have the $10 bills show 75% of the value of the $20 bills?

4. Now look at the $5 bills and the $10 bills. Do the $5 bills show 50% of the new value of the $10 bills?

   How can you change that?

5. Continue to compare the coins and bills, changing the amounts to fit the requirements.

6. After Allie solved her challenge, what coins and bills did she have left?

LOOK BACK
- Read the problem again.
- Check over your work.
- Did you answer the question that was asked?
- Does your answer make sense?

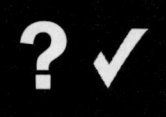

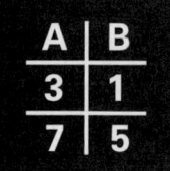

**Guess and Check
Use or Make a Table**

3 Jake, Mel, and Annie went mountain biking on Mammoth Mountain. Each biker took a different trail. They biked for a total of 10 hours and 10 minutes. Mel biked $\frac{2}{3}$ as long as Jake, and Annie biked 30 minutes less than Jake did. How long did each of them bike?

FIND OUT

What do you have to find out to solve the problem?

What does the problem tell you about Mel's time?

What does the problem tell you about Annie's time?

What do you know about how long they biked altogether?

CHOOSE STRATEGIES

You can **Guess and Check** and **Use or Make a Table** to help you solve this kind of problem. Guess numbers and record them in the table. Check to see if the total time is correct. If your guesses are not right, guess again. Use the information from wrong guesses to help you make better guesses.

SOLVE IT

Look at the table that has been started. Keep track of your guesses in the table.

Time for Jake (J)	Time for Mel ($M = \frac{2}{3} \times J$)	Time for Annie ($A = J - 30$ min)	Total time (10 hr 10 min)

1. Begin with a guess for Jake's time. What is your guess? Write this in the table.

2. Now how can you find the time for Mel?

   What number can you write in the table for Mel's time?

3. How can you find the time for Annie?

   What number can you write in the table for Annie's time?

4. How can you check your guesses?

   What is the total time for the three bikers?

5. Are your guesses right?

6. If your guesses are wrong, what is wrong about them?

   How can this information help you with your next guess?

7. Keep guessing and checking until you find the right total time for the three bikers.

8. How long did each of them bike?

LOOK BACK
- Read the problem again.
- Check over your work.
- Did you answer the question that was asked?
- Does your answer make sense?

Problem Solver II 7

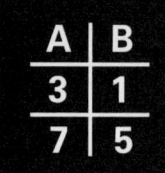

| ? | ✓ | A\|B
3\|1
7\|5 | **Guess and Check**
Use or Make a Table |

4 Della works at the Fish Palace. Today she is filling three fish tanks with water. Together the three tanks hold $29\frac{1}{4}$ gallons. Tank B holds half as much water as Tank A, and Tank C holds 8 quarts less than Tank A. How many gallons of water does each tank hold?

FIND OUT

What do you have to find out to solve the problem?

What do you know about Tank B?

What do you know about Tank C?

How much water do the three tanks hold altogether?

CHOOSE STRATEGIES

You can **Guess and Check** and **Use or Make a Table** to help you solve this kind of problem. Guess numbers and record them in the table. Check to see if the total amount is correct. If your guesses are not right, guess again. Use the information from wrong guesses to help you make better guesses.

8   Problem Solver II

SOLVE IT

Look at the table that has been started. Keep track of your guesses in the table.

Tank A (A)	Tank B $(B = \frac{1}{2}A)$	Tank C $(C = A - 2)$	Total $(29\frac{1}{4}$ gallons)

1. Begin with a guess for Tank A. What is your guess?

   Write this in the table.

2. How can you find the number of gallons for Tank B?

   What number can you write in the table for Tank B?

3. Look at the heading for Tank C. Why do you think it says A – 2 instead of A – 8?

   How can you find the number of gallons for Tank C?

   What number can you write in the table for Tank C?

4. How can you check your guesses?

   What is the total?

5. Are your guesses right?

6. If your guesses are wrong, what is wrong about them?

   How can this information help you with your next guess?

7. Keep guessing and checking until you find the right total number of gallons.

8. How many gallons of water does each tank hold?

LOOK BACK
- Read the problem again.
- Check over your work.
- Did you answer the question that was asked?
- Does your answer make sense?

A B C
Make It Simpler

5 In 2001, a hiker walked all three major U.S. hiking trails. He was the first person to do this in one calendar year. He averaged 158,400 feet per day. If he walked 7,371 miles in all, about how many months did it take him?

FIND OUT

What do you have to find out to solve the problem?

What does the problem tell you about the distance that he hiked?

CHOOSE STRATEGIES

You can **Make It Simpler** to help you solve the problem.
To make the problem simpler, start with finding out how many miles he went in one day.

SOLVE IT

1. What do you know about how far the hiker went in one day?

   How can you find out how many miles that is?

   How many miles did he go in one day?

2. How can you find out how many months it took him?

   How can you find out the number of days he hiked?

   How many days did he hike?

3. Now, how can you find the number of months that he hiked?

4. About how many months did it take him?

LOOK BACK
- Read the problem again.
- Check over your work.
- Did you answer the question that was asked?
- Does your answer make sense?

Problem Solver II 11

Make It Simpler

6 Four helicopter pilots are on fire duty. They each lower a helicopter over Bass Lake, dip a bucket into the lake, and fill it with 375 gallons of water. They fly to the fire, drop the water, then fly back to refill their buckets. On Monday they all made the same number of round trips. Together they carried 48 tons of water to the fire. How many round trips did each pilot make?

FIND OUT

What do you have to find out to solve the problem?

What does the problem tell you about the helicopter pilots?

What do you know about what the pilots did on Monday?

CHOOSE STRATEGIES

You can **Make It Simpler** to help you solve the problem. To make the problem simpler, first find out the number of tons that they carried for one trip.

12   Problem Solver II

SOLVE IT

1. What is one way you could make this problem simpler?

2. What do you know about the water a pilot carries on one round trip?

3. How can you convert gallons to tons?

   How can you find out how many pounds a gallon of water weighs?

   How many ounces are there in a pint?

   How many ounces are there in a quart?

   How many ounces are there in a gallon?

   How many pounds does a gallon of water weigh?

4. How many pounds are there in a ton?

   How can you find out how many gallons of water weigh a ton?

   How many gallons are in a ton?

5. If each pilot carries 375 gallons on a trip, then how can you find out how many tons that is?

   How many tons of water do they carry on one trip?

6. If you know the number of tons for one trip, then how can you find the number of trips needed for 48 tons?

   How many trips are needed?

7. How do you find the number of trips for each pilot?

8. How many round trips did each pilot make?

LOOK BACK
- Read the problem again.
- Check over your work.
- Did you answer the question that was asked?
- Does your answer make sense?

Problem Solver II 13

Make an Organized List
Act Out or Use Objects

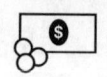

Play Money

7 One week Cathy, Victoria, and Elena earned $60.00 for their summer yard and pet business. They received the $60.00 in $5 bills, $10 bills, $20 bills, or some combination of those bills. If they were able to divide up the money evenly, what kinds of bills or combination of bills could they have had?

Find all the possible solutions.

FIND OUT

What do you have to find out to solve the problem?

What does the problem tell you about the total amount of money that they received?

What do you know about the kinds of bills they received?

What do you know about how they divided the money?

CHOOSE STRATEGIES

You can **Make an Organized List** and **Act Out or Use Objects** to help you solve this kind of problem. Use play money to help you think about the problem and work out a solution.

14   Problem Solver II

SOLVE IT

Look at the organized list that has been started.

$5 bills	$10 bills	$20 bills	Total
12			$60

1. What will you keep track of in the rows?

2. What is the total value for the first row?

   Is the total right?

   Is there another way you can use $5 bills with either $10 bills or $20 bills to make $60?

3. Try making other combinations of bills. Use the list to help organize your thinking.

4. What kinds of bills or combination of bills could they have had?

LOOK BACK
- Read the problem again.
- Check over your work.
- Did you answer the question that was asked?
- Does your answer make sense?

Make an Organized List

8 The Salad Green just opened, and the owners are pleased with their sales. On Friday they sold between 4 dozen and 6 dozen salads. They sold $\frac{1}{2}$ as many spinach as pasta salads. They sold $\frac{1}{3}$ as many spinach as potato salads. How many of each kind of salad could they have sold?

Find all the possible solutions.

FIND OUT
What do you have to find out to solve the problem?

What does the problem tell you about the total number of salads that they sold?

What do you know about how many of each kind of salad they sold?

CHOOSE STRATEGIES
You can **Make an Organized List** to help you solve this kind of problem.

SOLVE IT

Look at the organized list that has been started.

Spinach	Pasta	Potato	Total
5			

1. What will you keep track of in the rows?

2. If the number of spinach salads is 5, then what would the number of pasta salads be?

   Potato salads?

   What would the total number of salads be?

   Is this number between 4 dozen and 6 dozen?

3. Try other numbers for each kind of salad, until you find totals that are between 4 dozen and 6 dozen.

4. How many of each kind of salad could they have sold?

LOOK BACK
- Read the problem again.
- Check over your work.
- Did you answer the question that was asked?
- Does your answer make sense?

Problem Solver II 17

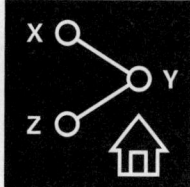

**Work Backwards
Use or Make a Picture or Diagram**

9 Jordan is working on his uncle's fishing boat. When the boat got back to port, they sold $\frac{1}{5}$ of their fish to the Nelson Fish Canning Company. Then they sold $\frac{3}{8}$ of what they had left to a restaurant. Finally, they sold $\frac{9}{10}$ of the remaining fish to the market. They had 130 fish left to take home. How many fish did Jordan and his uncle catch?

FIND OUT

What do you have to find out to solve the problem?

What does the problem tell you about what they did with the fish they caught?

What do you know about the number of fish they took home?

CHOOSE STRATEGIES

You can **Work Backwards** and **Use or Make a Picture or Diagram** to help you solve this kind of problem. Begin with the information at the end of the problem, then work backwards to find the missing numbers. Make a diagram to show the fractional amounts.

18 Problem Solver II

SOLVE IT

Look at the diagram that has been started.

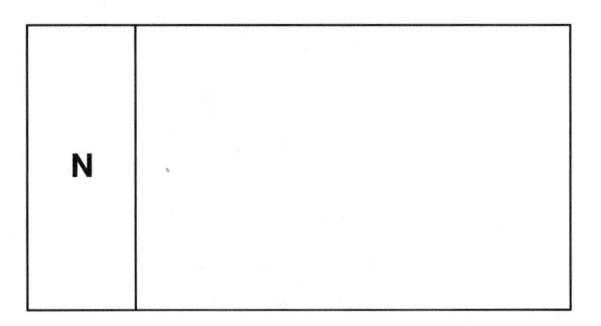

```
+-----+-----------+
|     |           |
|  N  |           |
|     |           |
+-----+-----------+
```

1. What does the diagram show now?

   Why is one space labeled N?

2. How should you divide the rest of the rectangle?

   How many of those equal parts should you label?

   How can you label them?

3. How should you divide the space that is not labeled?

   How many of those parts should you label?

   How can you label them?

4. What number does the problem give you, and in what fractional part of the diagram should you write it?

5. If you work backwards, then what amount can you write in each of the spaces labeled M for market?

6. Working backwards again, what numbers can you write in each of the other spaces?

7. How many fish did they catch?

LOOK BACK
- Read the problem again.
- Check over your work.
- Did you answer the question that was asked?
- Does your answer make sense?

Problem Solver II 19

Work Backwards

10 Darryl and Tom ran from the Tower of Trouble carrying a bag of gold chips. A blue dragon stopped them and demanded $\frac{1}{2}$ of their chips plus 3 more. Next, a giant snake blocked their way and asked for $\frac{1}{2}$ of the chips they had left plus 3 more. Then a knight said he would help them if they gave him $\frac{1}{2}$ of their remaining chips plus 3 more. They escaped with 3 gold chips. How many gold chips were in their bag when they met the dragon?

FIND OUT

What do you have to find out to solve the problem?

What does the problem tell you about how Darryl and Tom lost their gold chips?

How many chips did they escape with?

CHOOSE STRATEGIES

You can **Work Backwards** to help you solve this kind of problem. Begin with the information at the end of the problem, then work backwards to find the missing numbers.

SOLVE IT

1. Begin at the end of the problem. What is the last thing that happened?

   How many chips did they have left?

   You can write an equation to show this. Use C to represent the chips they had before the knight took some. What equation will show this?

2. How can you find the value of C before the knight?

3. Now work backwards. Write an equation to show what happened with the snake and how many chips they had left after the snake. What equation will show this?

4. How can you find the value of C before the snake?

5. Keep working backwards and writing equations.

6. How many gold chips were in their bag when they met the dragon?

LOOK BACK
- Read the problem again.
- Check over your work.
- Did you answer the question that was asked?
- Does your answer make sense?

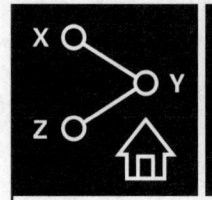

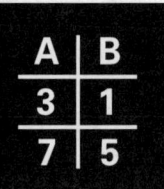

Use or Make a Picture or Diagram
Use or Make a Table
Use or Look for a Pattern

11 On the HG500, gears are grinding, pulleys are whirring, and wheels are turning. The graph shows how many times each wheel made a full rotation in the last 4 minutes. When wheel A had made 9 rotations, wheel B had made 15 rotations and wheel C had made 21 rotations. When wheel D had made 8 rotations, wheel E had made half that many.

The wheels continue to rotate at their original rate. When wheel C has made 42 rotations, how many rotations will each of the other wheels have made?

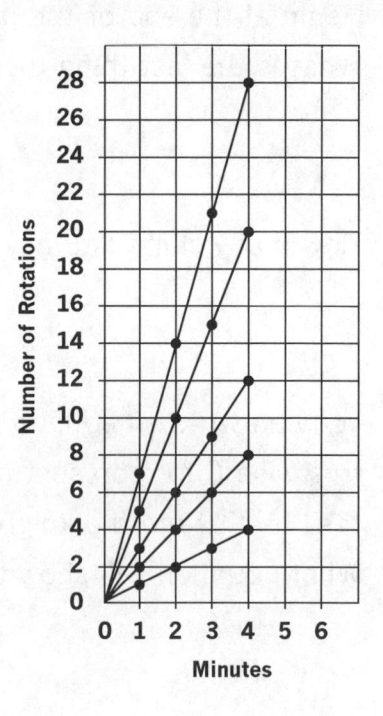

FIND OUT

What do you have to find out to solve the problem?

What does the graph show?

What do you know about wheels A, B, and C?

What do you know about wheels D and E?

CHOOSE STRATEGIES

You can **Use or Make a Picture or Diagram, Use or Make a Table,** and **Use or Look for a Pattern** to help you solve this kind of problem. Use the graph to find information. Make a table to keep track of the numbers of rotations made by the wheels. Look for patterns in the numbers.

22   Problem Solver II

SOLVE IT

Look at the graph and at the table that has been started.

Minutes	Number of Rotations for Each Wheel				
	A	B	C	D	E
1					
2					
3					

1. Find the line in the graph that shows a wheel that made 9 rotations. Label this line A. How many rotations did wheel A make in 1 minute?

   In 2 minutes?   In 3 minutes?

   Write those numbers in the table.

2. Find the line in the graph that shows a wheel that had made 15 rotations when wheel A had made 9 rotations. What label will you give this line?

   How many rotations did that wheel make in 1 minute?

   In 2 minutes?   In 3 minutes?

   Write those numbers in the table.

3. Keep using the graph and filling in the table. Look for patterns. Use the patterns to help you finish the table.

4. How many rotations will wheels A, B, D, and E have made when wheel C has made 42 rotations?

LOOK BACK
- Read the problem again.
- Check over your work.
- Did you answer the question that was asked?
- Does your answer make sense?

Problem Solver II   23

12 The Fantasy Shop rents out costumes. The graph shows the usual number of times that different costumes are rented per month. When the dinosaur costume has been rented 24 times, the gorilla costume has been rented half that many times, and the knight costume has been rented 28 times. When the monster costume has been rented 10 times, the pirate costume has been rented 16 times. If this pattern continues, when the dinosaur costume has been rented 36 times, how many times will the gorilla, knight, monster, and pirate costumes have been rented?

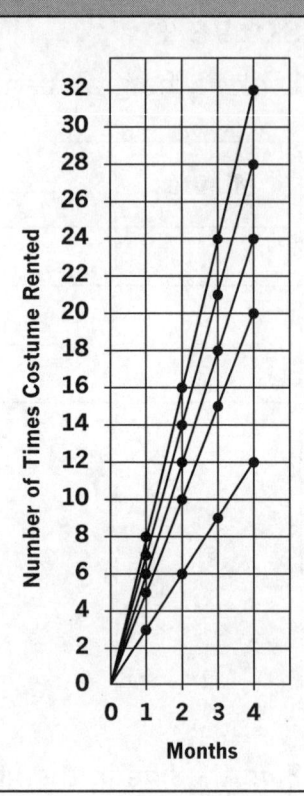

FIND OUT

What do you have to find out to solve the problem?

What does the graph show?

What do you know about the dinosaur, gorilla, and knight costumes?

What do you know about the monster and pirate costumes?

CHOOSE STRATEGIES

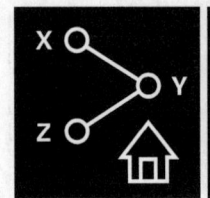

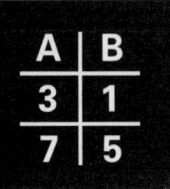

You can **Use or Make a Picture or Diagram, Use or Make a Table,** and **Use or Look for a Pattern** to help you solve this kind of problem. Use the graph to find information. Make a table to keep track of the numbers of times the costumes were rented. Look for patterns in the numbers.

24   Problem Solver II

SOLVE IT

Look at the graph and at the table that has been started.

Month	Number of Times Costume Rented				
	Dinosaur	Gorilla	Knight	Monster	Pirate
1					

1. Find the lines in the graph that show one costume that had been rented out 24 times and another costume that had been rented out half that many times in the same number of months. Which costumes do they represent?

   Label each line with the first letter of the name of the costume.

2. How many times had the dinosaur costume been rented out in 1 month?

   In 2 months?   In 3 months?

   Write those numbers in the table.

3. How many times had the gorilla costume been rented out in 1 month?

   In 2 months?   In 3 months?

   Write those numbers in the table.

4. Keep using the graph and filling in the table. Look for patterns. Use the patterns to help you finish the table.

5. How many times will the gorilla, knight, monster, and pirate costumes have been rented when the dinosaur costume has been rented 36 times?

LOOK BACK
- Read the problem again.
- Check over your work.
- Did you answer the question that was asked?
- Does your answer make sense?

13

Al, Benner, Denzel, Guido, Nan, Stacy, Francine, and Joy all go to Meadow School and live in the East Wood Apartments.

- The number of Francine's apartment is a palindrome. She lives directly below a boy's apartment and next door to a girl's apartment.
- Guido's apartment number is the highest prime number on any of these apartments. He lives next door to a boy and directly below a girl.
- Benner's apartment is between Nan's apartment and Stacy's apartment.
- The digits of Stacy's apartment number make the sum of 9.
- Denzel does not live directly above Francine.

In which apartment does each person live?

201	203	205	207
101	103	105	107

FIND OUT

What do you have to find out to solve the problem?

Who lives in these apartments?

What do the clues tell you?

CHOOSE STRATEGIES

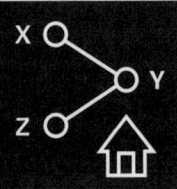

You can **Use Logical Reasoning, Use or Make a Picture or Diagram,** and **Act Out or Use Objects** to help you solve this kind of problem. Write each name on a small strip of paper. Move the paper strips around on the diagram until they fit the clues.

SOLVE IT

1. Read the first clue. Which number is a palindrome?

   Put the paper with Francine's name at this apartment. What do you know about Francine's neighbors?

   You can write G (for *girl*) and B (for *boy*) where they belong, and move papers there when you have more information.

2. Read the next clue. Which apartment numbers are prime numbers?

   Where can you put Guido's name?

   What do you know about Guido's neighbors?

   Write G and B where they belong.

3. Read the next clue. If Benner lives between two girls, then could he live on the first floor?

   Where could he live on the second floor?

   Put his name there.

4. Read the next clue. If you know where the girls live, then you can check to see which of their numbers has a sum of 9. Where can you put Stacy's name?

5. Read the last clue. If Denzel does not live directly above Francine, then where does he live?

6. In which apartment does each person live?

LOOK BACK
- Read the problem again.
- Check over your work.
- Did you answer the question that was asked?
- Does your answer make sense?

Problem Solver II   27

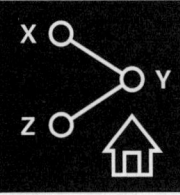

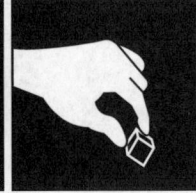

Use Logical Reasoning
Use or Make a Picture or Diagram
Act Out or Use Objects

14

Tamara, Benito, Ken, Stu, Lina, Alonso, Dee, and Mindy are sitting at a rectangular table with one person at each end. Each one has a different profession: rock singer, paramedic, doctor, film animator, computer programmer, forest ranger, ski instructor, and pilot.

- Lina is sitting at one end of the table. She is sitting between two men, neither of whom is her brother Stu, a film animator.
- The forest ranger is sitting at one end of the table with a man on her left and a woman on her right.
- The computer programmer is sitting between Alonso and Tamara, and he is directly across from Dee.
- The doctor is sitting to the left of Stu. On her left is the rock singer.
- Alonso is a ski instructor, and he is sitting between Benito and the paramedic.

Where is each person sitting at the table, and what work does each one do?

FIND OUT

What do you have to find out to solve the problem?

Who is sitting at the table?

What do you know about their professions?

What do the clues tell you?

CHOOSE STRATEGIES

You can **Use Logical Reasoning, Use or Make a Picture or Diagram,** and **Act Out or Use Objects** to help you solve this kind of problem. Write each name and each profession on a small strip of paper. Move the paper strips around on the diagram until they fit the clues.

SOLVE IT

1. Read the first clue. Where can you put Lina?

   Put her at the table. Who is she sitting next to?

   Write an M (for *man*) on each side of Lina.

2. Read the next clue. Where can you put the forest ranger?

   Is the forest ranger a man or woman?

   Who is the forest ranger sitting next to?

   Write M and W where they belong.

3. Read the next clue. If the computer programmer is sitting between a man and a woman, then where can you place the programmer?

   Is the programmer a man or a woman?

   Where can you put Alonso, Tamara, and Dee?

4. Keep using the clues and filling in the diagram.
5. Where is each person sitting, and what work does each one do?

LOOK BACK
- Read the problem again.
- Check over your work.
- Did you answer the question that was asked?
- Does your answer make sense?

Problem Solver II   29

Use Logical Reasoning
Make an Organized List

15

The Amazing Critter Shop is having its grand sale on dancing goats, singing pigeons, whistling mice, and laughing fish. The first customer paid $2,475 for 1 goat, 3 pigeons, and 2 mice. The second customer paid $1,525 for 3 pigeons and 2 mice. The third customer paid $5,725 for 3 goats and 5 mice. The fourth customer paid $1,580 for 1 mouse, 1 pigeon, and 4 fish. Same kinds of critters sold for the same price. Different kinds of critters sold for different prices. What was the price of each kind of amazing critter?

FIND OUT

What do you have to find out to solve the problem?

What does the problem tell you about the price of goats, pigeons, mice, and fish?

How much did each customer pay, and what did he or she buy?

CHOOSE STRATEGIES

You can **Use Logical Reasoning** and **Make an Organized List** to help you solve this kind of problem. Use "if ... then" logical thinking to find the price of the critters. **If** you know one thing is true, **then** you can find what else is true or not true. Make an organized list to keep track of the critters the customers bought and the amount they paid for them.

SOLVE IT

Look at the organized list that has been started.

Customer	Paid	Critters Bought
First		

1. Fill in the amounts the four customers paid, and list the critters they bought.

2. Look at what the first and second customers bought. How are their purchases alike and how are they different?

3. If you know what the first and the second customers paid, then how can you find out what 1 goat cost?

   What was the price for 1 goat?

4. If you know how much 1 goat cost, then how much did 3 goats cost?

5. If you know how much the third customer paid for 3 goats and 5 mice, and you know the cost of 3 goats, then how can you find out what 5 mice cost?

   What was the cost of 5 mice?

6. Keep using logical reasoning to help you find the price for each mouse, each pigeon, and each fish.

7. What was the price of each kind of amazing critter?

LOOK BACK
- Read the problem again.
- Check over your work.
- Did you answer the question that was asked?
- Does your answer make sense?

Problem Solver II   31

Use Logical Reasoning
Make an Organized List

16 Andy, Cody, Brad, and Dina each bought tickets at Action Attraction Park. Andy paid $16.25 for 1 Go-Kart, 1 Bumper Boat, and 1 Mini Golf ticket. Cody paid $21.00 for 1 Bumper Boat, 1 Mini Golf, and 2 Paddle Boat tickets. Brad paid $21.50 for 1 Go-Kart, 1 Bumper Boat, 1 Mini Golf, and 1 Paddle Boat ticket. Dina paid $22.00 for 1 Go-Kart, 1 Paddle Boat, and 2 Bumper Boat tickets. What was the cost of each kind of ticket?

FIND OUT
What do you have to find out to solve the problem?

What did each friend pay, and what tickets did he or she buy?

CHOOSE STRATEGIES
You can **Use Logical Reasoning** and **Make an Organized List** to help you solve this kind of problem. Use "if ... then" logical thinking to find the cost of the tickets. **If** you know one thing is true, **then** you can find what else is true or not true. Make an organized list to keep track of the tickets the friends bought and the amounts they paid.

32 Problem Solver II

SOLVE IT

Look at the organized list that has been started.

Person	Paid	Tickets Bought
Andy		
Cody		

1. Fill in the amounts the four friends paid, and list the tickets they bought.

2. Look at the tickets that Andy bought. Did any of the other friends buy similar tickets?

   How were their tickets alike and how were they different?

3. If you know what Andy paid for his tickets, and you know what Brad paid for his tickets, then how can you find out what 1 Paddle Boat ticket cost?

What was the price for 1 Paddle Boat ticket?

4. Then how much did 2 Paddle Boat tickets cost?

5. If you know how much Cody paid for his tickets, and you know the cost of 2 Paddle Boat tickets, then how much did 1 Bumper Boat ticket and 1 Mini Golf ticket cost?

6. Keep using logical reasoning to help you find the cost of each Go-Kart ticket, each Bumper Boat ticket, and each Mini Golf ticket.

7. What was the cost of each kind of ticket?

LOOK BACK
- Read the problem again.
- Check over your work.
- Did you answer the question that was asked?
- Does your answer make sense?

Problem Solver II 33

Use Logical Reasoning
Make an Organized List

17 Rosalina is in charge of supplies for the Children's Theater. The boxes on the left side of the scale balance the boxes on the right side of the scale. Together all six boxes weigh 1 kilogram.

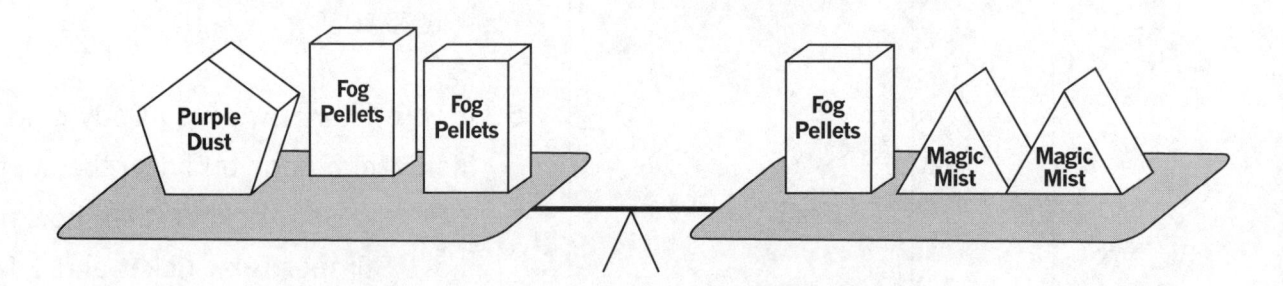

Boxes with the same shape weigh the same amount. Boxes with different shapes weigh different amounts. Each box weighs a whole number of grams between 99 and 201, and the number is divisible by 5. What could each box weigh?

Find all the possible solutions.

FIND OUT

What do you have to find out to solve the problem?

What does the problem tell you about the weight of the boxes on both sides of the scale?

What is the total weight of the boxes?

What else does the problem tell you about the weight of the boxes?

CHOOSE STRATEGIES

You can **Use Logical Reasoning** and **Make an Organized List** to help you solve this kind of problem. Use "if ... then" logical thinking. Make an organized list to help you find all possible weights of the boxes.

SOLVE IT

Look at the organized list that has been started.

Possible Weights in Grams

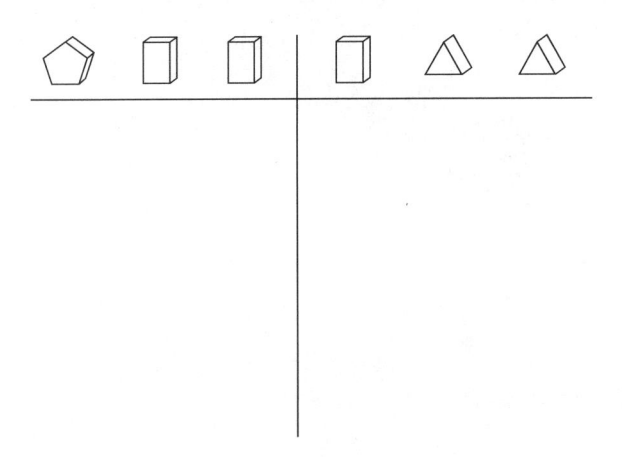

⬠ ▯ ▯ | ▯ △ △

1. If the total weight of the boxes is 1 kilogram, how many grams do the boxes on each side of the scale weigh?

2. What is the least number of grams a box of Purple Dust could weigh?

   Why?

If a box of Purple Dust weighs that number of grams, how many grams would the two boxes of Fog Pellets weigh together?

How many grams would each box of Fog Pellets weigh?

Then how much would each box of Magic Mist weigh?

Write those numbers in the organized list.

3. Could the box of Purple Dust weigh 105 grams?

4. Keep filling in the organized list until you find all possible weights for each box.

5. What are all the amounts each box could weigh?

LOOK BACK

- Read the problem again.
- Check over your work.
- Did you answer the question that was asked?
- Does your answer make sense?

Problem Solver II 35

**Use Logical Reasoning
Make an Organized List**

18 The reservations clerk at Ponderosa Park needs to place 790 people in campsites. Half of the people will be placed in Moonbeam campground, and half will be in the Riverside campground. Moonbeam has group campsites A1, A2, and B. Riverside has group campsites B, C1, C2, and C3. Campsites named with the same letter will have the same number of people. Campsites named with different letters will have different numbers of people. Each campsite will have between 79 and 151 people. What are all the possible numbers of people that could be in each group campsite?

FIND OUT

What do you have to find out to solve the problem?

What is the total number of people that need camping spaces?

How many people will be in Moonbeam campground?

How many people will be in Riverside campground?

What else does the problem tell you about the campgrounds?

Problem Solver II

CHOOSE STRATEGIES

You can **Use Logical Reasoning** and **Make an Organized List** to help you solve this kind of problem. Use "if ... then" logical thinking. Make an organized list to help you find all possible numbers of people in each group campsite.

SOLVE IT

Look at the organized list that has been started.

Possible Numbers of People in Each Campsite

A1	A2	B	B	C1	C2	C3

1. If the total number of people is 790, how many people will be in each campground?

2. Could there be 80 people in campsite A1 and campsite A2 in Moonbeam campground?

3. What is the least number of people there could be in campsite A1 and campsite A2?

   Then how many people would be in campsite B?

   Write those numbers in the list.

4. How many people would be in campsite B in Riverside campground?

   Then how many people would be in each of the three C campsites?

   Write those numbers in the list.

5. Does this make a total of 395 people in each campground?

6. Keep filling in the list until you find all possible numbers of people for each campsite.

7. What are all the different numbers of people that could be in each campsite?

LOOK BACK
- Read the problem again.
- Check over your work.
- Did you answer the question that was asked?
- Does your answer make sense?

Problem Solver II   37

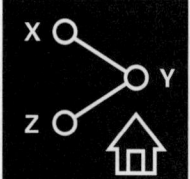

Use or Make a Picture or Diagram
Use Logical Reasoning

19

The principal of Lee School took a survey of 110 fifth and sixth graders to see what they did over the summer. She found 20 who went to sports camp, 30 who went to summer school, and 45 who went to science camp. Twelve students went only to sports camp, and 4 students went to sports camp and to science camp. How many of the students surveyed didn't do any of the three activities?

FIND OUT

What do you have to find out to solve the problem?

What does the problem tell you about who was surveyed?

What do you know about the student responses?

CHOOSE STRATEGIES

You can **Use or Make a Picture or Diagram** and **Use Logical Reasoning** to help you solve this kind of problem. Use a Venn circle diagram. In this problem, the diagram has three circles that overlap.

SOLVE IT

1. Look at the Venn circle diagram.

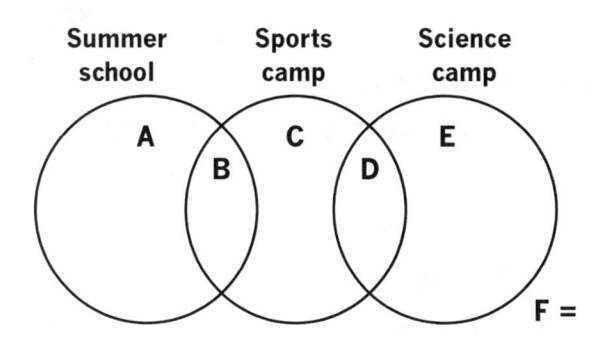

   Summer school — Sports camp — Science camp
   A, B, C, D, E
   F =

   What is the label at the top of the first circle?

   At the top of the second circle?

   At the top of the third circle?

2. A space where the two circles overlap is inside two circles. What label could you give space B?

3. What label could you give space D?

   How many students went to sports camp and to science camp?

   Write that number in space D.

4. How many students went to science camp?

   Does that include the students who went to sports camp?

   Then how can you find out how many students went only to science camp?

   Write that number in space E.

5. How many students went only to sports camp?

   Write that number where it belongs.

6. How many students altogether went to sports camp?

7. How can you find the number of students who went to summer school and to sports camp?

   Write that number in space B.

8. How many students went only to summer school?

9. How can you find the total number of students who did one or more of these activities?

   How many students did one or more activity?

10. How many of the students surveyed didn't do any of the three activities?

LOOK BACK

- Read the problem again.
- Check over your work.
- Did you answer the question that was asked?
- Does your answer make sense?

Problem Solver II

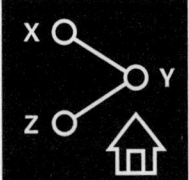

**Use or Make a Picture or Diagram
Use Logical Reasoning**

20

When the Hot Spot opened Saturday, the Grand Opening specials were burgers, taco salads, and wraps. Sixty people ordered burgers, 50 people ordered taco salads, and 44 people ordered wraps. Eight people ordered burgers and taco salads, 11 people ordered taco salads and wraps, and 40 people ordered only a burger. Three people ordered a burger, a taco salad, and a wrap. Then there were 18 people who ordered something other than the specials. How many people ordered something at the Hot Spot on Saturday?

FIND OUT

What do you have to find out to solve the problem?

What does the problem tell you about the people who came to the Hot Spot on Saturday?

CHOOSE STRATEGIES

You can **Use or Make a Picture or Diagram** and **Use Logical Reasoning** to help you solve this kind of problem. Use a Venn circle diagram. In this problem, the diagram has three circles that overlap.

40 Problem Solver II

SOLVE IT

1. Look at the Venn circle diagram.

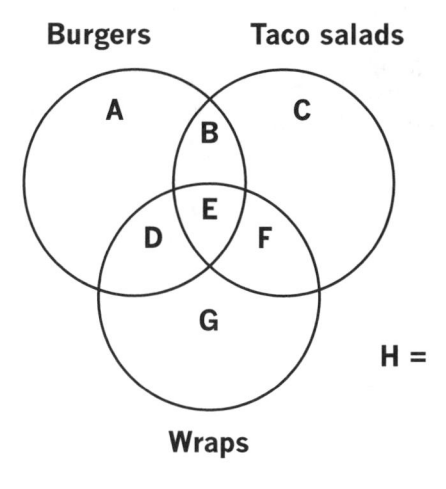

   Burgers   **Taco salads**

   A   B   C
   D   E   F
     G
   H =

   Wraps

   A space where two circles overlap is inside two circles. What label could you put on space B?

   On space D?

   On space F?

2. A space where the three circles overlap is inside three circles. What label could you put on space E?

   What number can you put in space E?

3. Do the 8 people who bought burgers and taco salads include the number who bought all three items?

   How can you find the number who bought just burgers and taco salads?

   What number can you put in space B?

4. Do the 11 people who bought taco salads and wraps include the 3 people who bought all three items?

   Find the number for space F and write it on the diagram.

5. If you have the numbers for spaces B, E, and F, then how can you find the number for space C, the people who only bought taco salads?

   What number can you put in space C?

6. Use the same reasoning to find the numbers for all the spaces of the circle labeled *Wraps*. Then find all the numbers for the *Burgers* circle.

7. What number belongs outside the circles in space H?

8. How many people ordered something at the Hot Spot on Saturday?

LOOK BACK
- Read the problem again.
- Check over your work.
- Did you answer the question that was asked?
- Does your answer make sense?

Problem Solver II   41

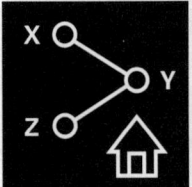

Use or Make a Picture or Diagram
Act Out or Use Objects

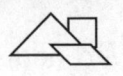

Tangrams

21 Luke took two pieces from a set of tangrams and made a quadrilateral. Then he moved the pieces around and made a different quadrilateral. The diagonals of the first shape were the same length and were perpendicular. The diagonals of the second shape were not the same length and were not perpendicular.

Which two pieces could Luke have used? What could his quadrilaterals and their diagonals have looked like?

More than one solution is possible.

FIND OUT
What do you have to find out to solve the problem?

What do you know about the pieces Luke used?

What do you know about the shapes Luke made?

CHOOSE STRATEGIES
You can **Act Out or Use Objects** and **Use or Make a Picture or Diagram** to help you solve this kind of problem. Use tangram pieces to make the shapes.

42   Problem Solver II

SOLVE IT

1. Which two pieces will you try first?

2. Can you put the two pieces together to make a quadrilateral?

3. Can you put the two pieces together to make a different quadrilateral?

4. Are the diagonals in one quadrilateral the same length and perpendicular?

5. Are the diagonals in the other quadrilateral different lengths and not perpendicular?

6. If you cannot make Luke's quadrilaterals with your two pieces, try a different set of two pieces.

7. Which two pieces could Luke have used?

   What could his quadrilaterals and their diagonals have looked like?

LOOK BACK

- Read the problem again.
- Check over your work.
- Did you answer the question that was asked?
- Does your answer make sense?

Problem Solver II 43

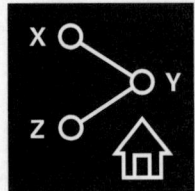

Use or Make a Picture or Diagram
Act Out or Use Objects

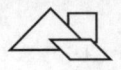

Tangrams

22 Kim, Shelly, and Amber have square gardens built around a fountain. The base of the fountain is a right triangle.

Kim's garden and Amber's garden each have 9 square yards of area. How many square yards of area does Shelly's garden have?

Trace tangram pieces to show how you got your answer.

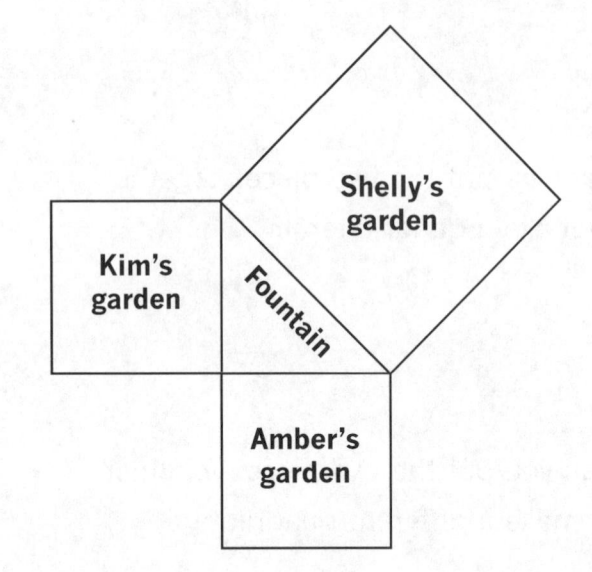

FIND OUT

What do you have to find out to solve the problem?

What does the problem tell you about the gardens of Kim, Amber, and Shelly?

What does the problem tell you about the fountain?

CHOOSE STRATEGIES

You can **Act Out or Use Objects** and **Use or Make a Picture or Diagram** to help you solve this kind of problem. Use tangram pieces to make the shapes.

SOLVE IT

1. Which tangram piece can you use to show the base of the fountain?

2. Which tangram pieces can you use to show Kim's and Amber's gardens?

3. How many square yards of area does each tangram square represent?

4. Use tangram pieces to represent Shelly's garden.

5. How many square yards of area does Shelly's garden have?

Trace tangram pieces to show how you got your answer.

LOOK BACK
- Read the problem again.
- Check over your work.
- Did you answer the question that was asked?
- Does your answer make sense?

Problem Solver II   45

23

Malcolm is solving a puzzle in a math contest. He looks at the first three squares in a series drawn on a dot array.

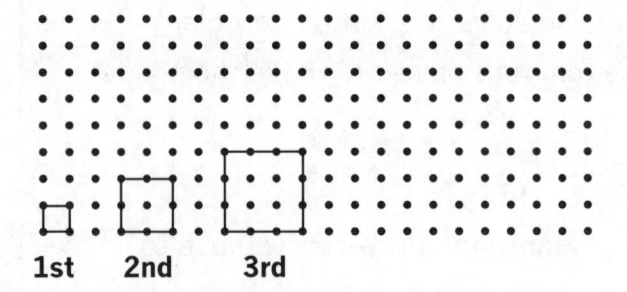

The first square touches 4 dots and has 1 square unit of area. The second square touches 8 dots and has 4 square units of area. Malcolm looks for a pattern in the way the squares get larger. His puzzle is to identify which square in the series will have $2\frac{1}{2}$ times as many square units of area as dots touched. What is the solution to Malcolm's puzzle?

FIND OUT

What do you have to find out to solve the problem?

What does the problem tell you about the squares Malcolm is looking at?

What does Malcolm look for in the way the squares get larger?

CHOOSE STRATEGIES

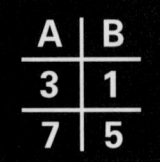

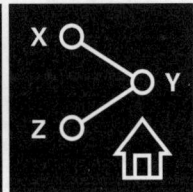

You can **Use or Make a Table, Use or Look for a Pattern,** and **Use or Make a Picture or Diagram** to help you solve this kind of problem. Use the diagram to see how the squares in the series get larger. Make a table to keep track of the numbers of dots touched and square units of area. Look for patterns in the numbers.

46 Problem Solver II

SOLVE IT

Look at the table that has been started.

Square	Number of Dots Touched	Square Units of Area
1st	4	1
2nd	8	4

1. How many dots are touched by the first square?

   How many square units of area are inside of it?

2. How many dots are touched by the second square?

   How many square units of area are inside of it?

3. How many dots are touched by the third square?

   How many square units of area are inside of it?

   Write those numbers in the table.

4. Look at the numbers in your table. What is the difference between the numbers of dots touched in the first and second squares?

   What is the difference between the numbers of dots touched in the second and third squares?

   What pattern do you see in the numbers of dots touched?

5. What is the difference between the areas of the first and second squares?

   Between the areas of the second and third squares?

   What pattern do you see in the numbers of square units of area?

6. Keep drawing more squares or use the patterns to fill in the rest of the table.

7. Which square in the series has $2\frac{1}{2}$ times as many square units of area as dots touched?

LOOK BACK
- Read the problem again.
- Check over your work.
- Did you answer the question that was asked?
- Does your answer make sense?

Problem Solver II   47

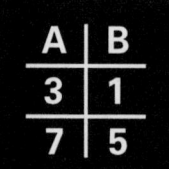

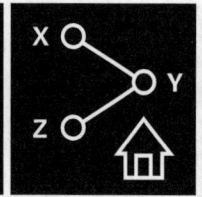

Use or Make a Table
Use or Look for a Pattern
Use or Make a Picture or Diagram

24

Tori likes to make up puzzles for her father to solve. She drew three rectangles on a dot array. The first rectangle had 2 square units of area and no dots inside. The second rectangle had 6 square units of area and 2 dots inside.

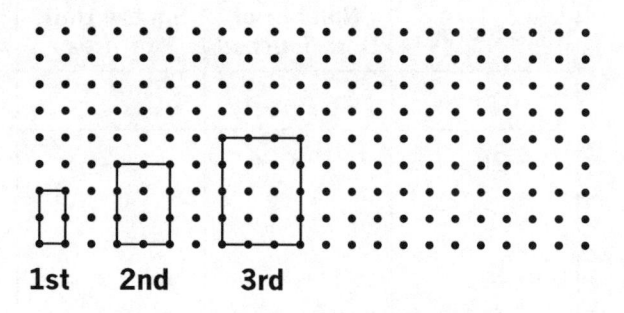

1st   2nd   3rd

Then Tori said, "Dad, if I keep using the same pattern to make larger rectangles in this series, which rectangle will have 20 more square units of area than dots inside?" How did Tori's dad solve her puzzle?

FIND OUT

What do you have to find out to solve the problem?

What does the problem tell you about the rectangles Tori drew?

CHOOSE STRATEGIES

You can **Use or Make a Table, Use or Look for a Pattern,** and **Use or Make a Picture or Diagram** to help you solve this kind of problem. Use the diagram to see how the rectangles in the series get larger. Make a table to keep track of the numbers of dots inside and square units of area. Look for patterns in the numbers.

48   Problem Solver II

SOLVE IT

Look at the table that has been started.

Rectangle	Number of Dots Inside	Square Units of Area
1st	0	2
2nd	2	6
3rd		

1. How many dots are inside the first rectangle?

   How many square units of area does it have?

2. How many dots are inside the second rectangle?

   How many square units of area does it have?

3. How many dots are inside the third rectangle?

   How many square units of area does it have?

   Write those numbers in the table.

4. Look at the numbers in your table. What is the difference between the numbers of dots inside the first and second rectangles?

   What is the difference between the numbers of dots inside the second and third rectangles?

   What pattern do you see in the numbers of dots?

5. What is the difference between the areas of the first and second rectangles?

   Between the areas of the second and third rectangles?

   What pattern do you see in the numbers of square units of area?

6. Keep drawing more rectangles or use the patterns to fill in the table.

7. Which rectangle in the series has 20 more square units of area than dots inside?

LOOK BACK
- Read the problem again.
- Check over your work.
- Did you answer the question that was asked?
- Does your answer make sense?

Problem Solver II   49

Brainstorm

25 Dave made up these three code puzzles for Andrew to solve:

9 S i T-T-T   5 i a R f B   64 S o a C

What were Andrew's solutions to the code puzzles?

FIND OUT

What do you have to find out to solve the problem?

What does the problem tell you about Dave's code puzzles?

CHOOSE STRATEGIES

You can **Brainstorm** to help you solve this kind of problem. Try to think about things in different ways than you usually think about them. Try out as many different ideas as you can. The answer may pop into your mind.

SOLVE IT

1. What does each code puzzle begin with?

2. Look at the first puzzle. What are some ways the number 9 might be used?

What about 5?

What about 64?

3. What do you think the letters might mean in the codes?

4. Keep brainstorming until you figure out what each code means.

5. What does each code mean?

LOOK BACK
- Read the problem again.
- Check over your work.
- Did you answer the question that was asked?
- Does your answer make sense?

Brainstorm

26 How can you add fewer than four lines and change this drawing of a 2-dimensional shape into a drawing of a 3-dimensional shape?

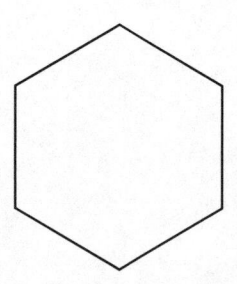

FIND OUT

What do you have to find out to solve the problem?

What does the problem tell you about how to change the shape?

CHOOSE STRATEGIES

You can **Brainstorm** to help you solve this kind of problem. Try to think about things in different ways than you usually think about them. Try out as many different ideas as you can. The answer may pop into your mind.

SOLVE IT

1. What kind of shape is shown?

How many sides does it have?

2. How many lines can you add?

Can you add these lines inside or outside of the shape?

3. Try adding lines in different ways until your drawing shows a 3-dimensional shape.

4. How can you add fewer than four lines and change this drawing of a 2-dimensional shape into a drawing of a 3-dimensional shape?

LOOK BACK
- Read the problem again.
- Check over your work.
- Did you answer the question that was asked?
- Does your answer make sense?

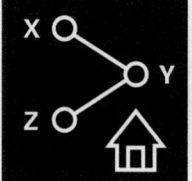

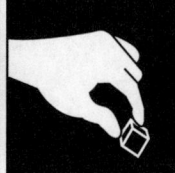

**Use or Make a Picture or Diagram
Act Out or Use Objects**

Cubes

27

This sketch shows 20 cube-shaped cartons stacked up on the floor of a warehouse.

This sketch shows a view looking down at the top of the stacked cartons.

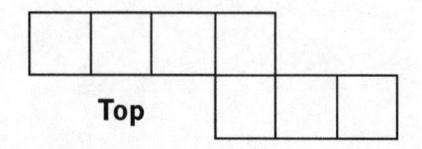

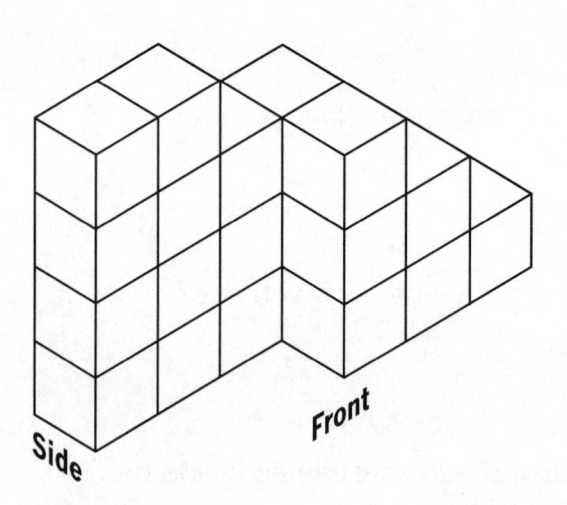

Top

What would views of the side and the front of the stacked cartons look like? How many square units of exposed surface area are there on the stacked cartons?

FIND OUT

What do you have to find out to solve the problem?

What does the problem tell you about the cartons?

54   Problem Solver II

CHOOSE STRATEGIES

You can **Act Out or Use Objects** and **Use or Make a Picture or Diagram** to help you solve this kind of problem. Look at the sketches and use cubes to model the stacked cartons.

SOLVE IT

1. How many cubes will you use to model the stacked cartons?

2. How many cubes does the top view of the stacked cartons show?

3. Build a model of the stacked cartons.

4. How many cubes do you see when you view the side of your model?

5. How many cubes do you see when you view the front of your model?

6. What would views of the side and the front of the stacked cartons look like?

How many square units of exposed surface area are there on the stacked cartons?

LOOK BACK
- Read the problem again.
- Check over your work.
- Did you answer the question that was asked?
- Does your answer make sense?

Problem Solver II 55

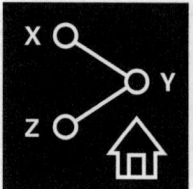

Use or Make a Picture or Diagram
Act Out or Use Objects

Cubes

28 The architect used 18 cubes to make a model of a building. This is her sketch of the model.

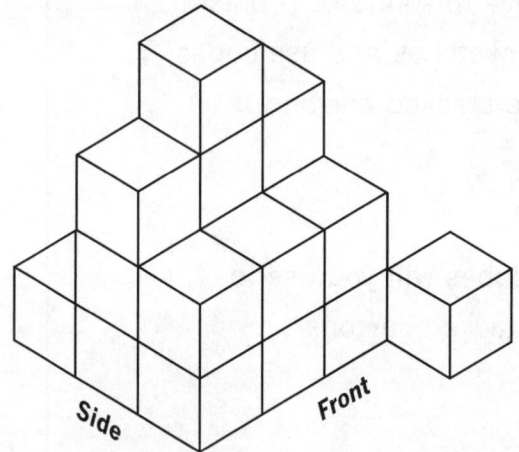

She also sketched her views of the side, the front, and the top of the model. What did her sketches look like? How many square units of exposed surface area are there on the model of the building?

FIND OUT

What do you have to find out to solve the problem?

What does the problem tell you the architect did?

56 Problem Solver II

CHOOSE STRATEGIES

You can **Act Out or Use Objects** and **Use or Make a Picture or Diagram** to help you solve this kind of problem. Look at the sketch and use cubes to build the model.

SOLVE IT

1. How many cubes will you use to build the model?

2. Build the model.

3. How many cubes do you see when you view the top of the model?

4. How many cubes do you see when you view the side of the model?

5. How many cubes do you see when you view the front of the model?

6. What did the architect's sketches of the top, front, and side of the model look like?

How many square units of exposed surface area are there on the model of the building?

LOOK BACK
- Read the problem again.
- Check over your work.
- Did you answer the question that was asked?
- Does your answer make sense?

Problem Solver II 57

Use Logical Reasoning
Act Out or Use Objects

Tangrams

29

Antonio is playing the game Mystery Shape.
He is using one set of tangram pieces.
He draws a card worth 5 points. The card says:

- Make a shape that has two angles greater than 90 degrees and two angles less than 90 degrees.
- The area of the tangram square is $\frac{1}{8}$ the area of the whole shape.

What shape could Antonio make to earn 5 points?
What would be the measure of each angle in the shape?

More than one solution is possible.

FIND OUT

What do you have to find out to solve the problem?

What does the problem tell you about the angles in the shape?

What do you know about the area of the shape?

CHOOSE STRATEGIES

You can **Use Logical Reasoning** and **Act Out or Use Objects** to help you solve this kind of problem. Use tangram pieces to make the shape.

58   Problem Solver II

SOLVE IT

1. Which of the tangram pieces have 90-degree angles?

2. How can you find angles less than and greater than 90 degrees by using the tangram pieces?

   Which pieces have angles greater than 90 degrees?

   Which pieces have angles less than 90 degrees?

3. If the square has an area of 1 unit, then what other shapes have an area of 1 unit?

   What pieces put together have an area of 1 unit?

   How many units of area does the large triangle have?

4. How many units of area does Antonio's shape need to have?

5. Try putting together different groups of tangram pieces until you make a shape that fits all the clues.

6. What shape could Antonio make to earn 5 points? What would be the measure of each angle in the shape?

. . . .

. . . .

. . . .

. . . .

. . . .

. . . .

LOOK BACK
- Read the problem again.
- Check over your work.
- Did you answer the question that was asked?
- Does your answer make sense?

Problem Solver II   59

Use Logical Reasoning
Act Out or Use Objects

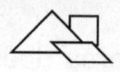

Tangrams

30

Josefina is playing What's My Measure? She has one set of tangram pieces. She draws a card that says:

- Make two similar shapes. Use at least three tangram pieces.
- The sum of the angles in each shape is 180 degrees.
- The area of one shape is four times the area of the other shape.

Let's say that the small triangle has 1 unit of area. What two shapes could Josefina make? How many units of area does each shape have?

FIND OUT

What do you have to find out to solve the problem?

What does the problem tell you about the shapes?

CHOOSE STRATEGIES

You can **Use Logical Reasoning** and **Act Out or Use Objects** to help you solve this kind of problem. Use tangram pieces to make the shapes.

60   Problem Solver II

SOLVE IT

1. Try making some similar shapes. Which tangram pieces did you use?

What shapes did you make?

Are the angles congruent?

Are the lengths of the sides proportional?

Is the sum of the angles of each shape 180 degrees?

2. If your shapes do not match all the clues, use other pieces and try making more similar shapes.

3. What two shapes could Josefina make?

How many units of area does each shape have?

LOOK BACK
- Read the problem again.
- Check over your work.
- Did you answer the question that was asked?
- Does your answer make sense?

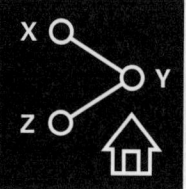

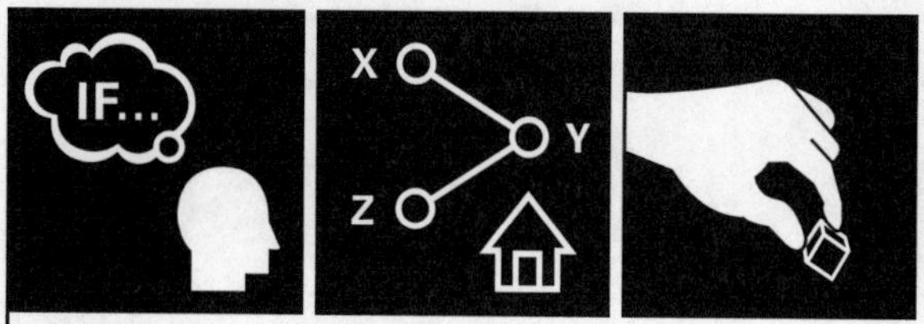

Use Logical Reasoning
Use or Make a Picture or Diagram
Act Out or Use Objects

3-D Shapes

31 Jud drew three concentric circles and gave each circle a label as shown. He gave a set of 3-D shapes to Paul and asked him to put the shapes where they belong in his circle diagram. He said that if a shape does not belong in the circles, it goes outside the circles.

Where should Paul put each of these 3-D shapes in Jud's diagram?

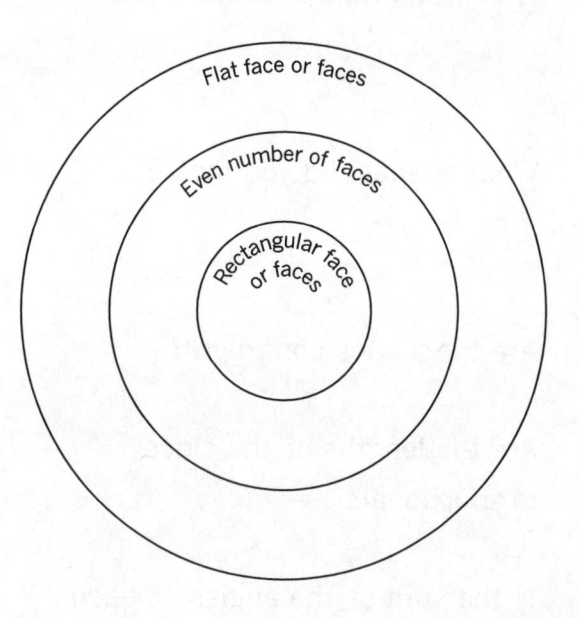

Flat face or faces
Even number of faces
Rectangular face or faces

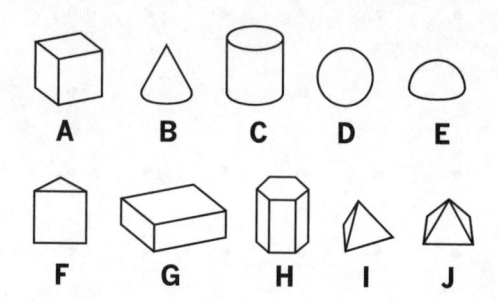

A B C D E
F G H I J

FIND OUT

What do you have to find out to solve the problem?

What are the labels in the circles, moving from the outside into the center?

Where do shapes go that do not belong in the circles?

CHOOSE STRATEGIES
You can **Use Logical Reasoning, Use or Make a Picture or Diagram,** and **Act Out or Use Objects** to help you solve this kind of problem. Use 3-D shapes and "if ... then" logical thinking to help you find where each shape belongs in the diagram.

SOLVE IT

1. Does shape A, the cube, have a flat face or faces?

   Does it have an even number of faces?

   Does it have a rectangular face or faces?

   Where does it belong?

   Write the letter A where it belongs.

2. Does shape B, the cone, have a flat face or faces?

   Does it have an even number of faces?

   Does it have a rectangular face or faces?

   Where does it belong?

   Write the letter B where it belongs.

3. Keep finding where the other 3-D shapes belong. Write their letters where they belong in the diagram.

4. Where should Paul put each 3-D shape in Jud's diagram?

LOOK BACK
- Read the problem again.
- Check over your work.
- Did you answer the question that was asked?
- Does your answer make sense?

Problem Solver II   63

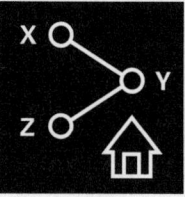

Use Logical Reasoning
Use or Make a Picture or Diagram
Act Out or Use Objects

3-D Shapes

32 Nya drew three concentric circles and put 3-D shapes inside and outside the circles as shown. What is Nya's rule for each circle? Where do the rest of the 3-D shapes belong in her diagram?

Write the label for each circle, and write the letters of shapes A, B, C, and D in the diagram where they belong.

More than one solution is possible.

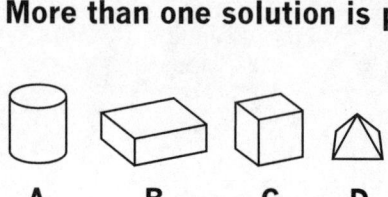

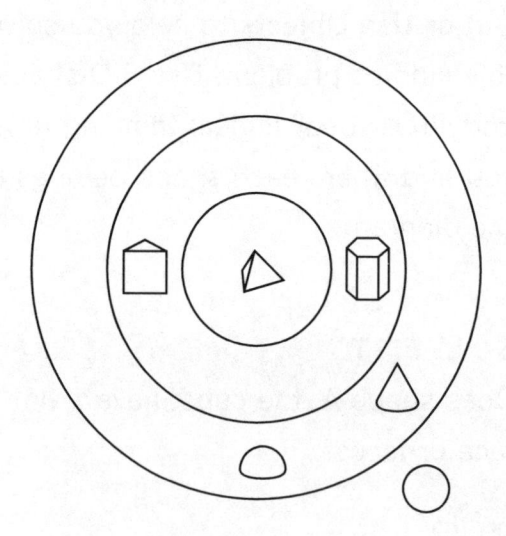

A   B   C   D

FIND OUT

What do you have to find out to solve the problem?

What 3-D shape did Nya put in the center circle?

What 3-D shapes did she put in the next larger circle?

What 3-D shapes did she put in the outer circle?

What 3-D shape did she put outside the circles?

CHOOSE STRATEGIES

You can **Use Logical Reasoning, Use or Make a Picture or Diagram,** and **Act Out or Use Objects** to help you solve this kind of problem. Use 3-D shapes and "if ... then" logical thinking to help you find where each shape belongs in the diagram.

SOLVE IT

1. In what way are the shapes in the circles different from the shape outside the circles?

2. What label could you give the outer circle?

3. Could shapes A, B, C, or D belong in that circle?

4. In what way are the shapes in the middle circle alike?

   How are they different from the shapes in the outer circle?

5. What label could you give the middle circle?

6. Could you move any of shapes A, B, C, or D into that circle?

7. In what way is the shape in the center circle different from the shapes shown in the middle circle?

8. What label could you give the center circle?

9. Where do shapes A, B, C, and D belong in the diagram? Write their letters to show where they belong. Write the labels for the circles.

LOOK BACK
- Read the problem again.
- Check over your work.
- Did you answer the question that was asked?
- Does your answer make sense?

Problem Solver II 65

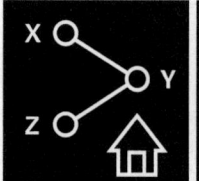

Use or Make a Picture or Diagram
Use Logical Reasoning

33

Card games, board games, and computer games sell well at The Game Factory.

- Carlos bought the kind of game that sold more than 400 units for 5 months.
- Derrick bought the kind that sold fewer than 250 units during 8 months.
- Irina bought the kind that sold 50 fewer units than Latoya's kind for two months in a row.
- Computer games sold the most units over 12 months, but in one month they sold 50 fewer units than card games and 100 fewer units than board games.

Which kind of game did Carlos, Irina, Derrick, and Latoya buy? How many units did each kind sell in all?

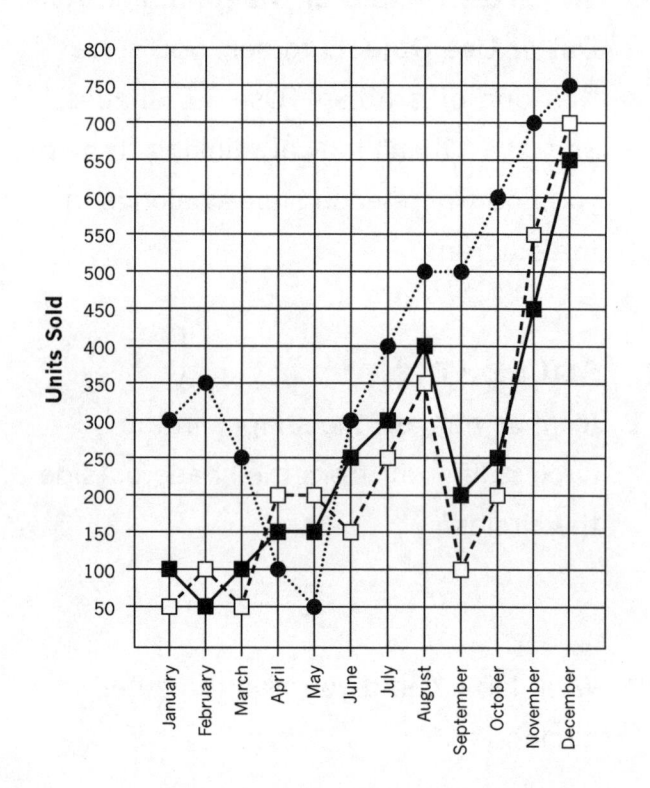

FIND OUT

What do you have to find out to solve the problem?

What does the problem tell you about the games that The Game Factory is selling?

What do you know about the game each person bought?

66 Problem Solver II

CHOOSE STRATEGIES

You can **Use or Make a Picture or Diagram** and **Use Logical Reasoning** to help you solve this kind of problem.

SOLVE IT

Use the graph and the clues to solve the problem.

1. Start by finding out which kind of game each line represents. Which hint can help you do this?

   How can you find which line shows computer games?

   Which line shows the computer games?

2. Can you use the same hint to find the lines for card games and board games? Why or why not?

   Which line is board games?

   Which line is card games?

3. Now use the clue for Carlos. What are you looking for in the graph?

   Which kind of game did Carlos buy?

4. Use the clue for Derrick. What are you looking for in the graph?

   Which kind of game did Derrick buy?

5. Read the clue for Irina and Latoya. Find the kind of game that they bought.

6. Which kind of game did Carlos, Irina, Derrick, and Latoya buy?

   How many units did each kind of game sell in all?

LOOK BACK
- Read the problem again.
- Check over your work.
- Did you answer the question that was asked?
- Does your answer make sense?

Problem Solver II 67

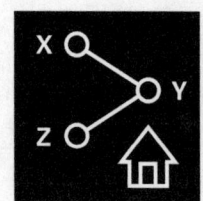

Use or Make a Picture or Diagram

34 The sixth and seventh graders estimated the amount of time they spend using a computer each week, rounded to the nearest quarter hour. The results are shown in the stem-and-leaf plot.

- Danielle uses a computer for 2 hours more than the median for sixth grade.
- Nadia uses a computer for 2 hours and 45 minutes less than the median for seventh grade.
- Taylor uses a computer for twice as long as the mode for the sixth grade.
- Katie uses a computer for 1 hour and 30 minutes more than the mode for seventh grade.

How much time do Danielle, Nadia, Taylor, and Katie each spend using a computer during the week?

Hours on the Computer

Sixth grade					Seventh grade				
.75	.5	.5	.25	0	.5	.5	.75	.75	
.75	.5	.0	.0	1	.0	.25	.5	.5	.5
	.5	.25	.0	.0	2	.0	.5	.5	.75
	.5	.0	.0	.0	3	.0	.5	.5	.75
	.5	.5	.0	.0	4	.0	.0	.25	.5
		.5	.25	.0	5	.0	.0	.5	.5
		.5	.25	.0	6	.0	.25	.5	
			.5	.0	7	.0	.5	.75	
					8	.0	.5	.5	
					9	.0	.25		

Key

.5 | 1 | .25

Sixth grade: Read from the middle to the left: 1.5

Seventh grade: Read from the middle to the right: 1.25

FIND OUT

What do you have to find out to solve the problem?

What does the problem tell you about the graph?

What do you know about each student?

68   Problem Solver II

CHOOSE STRATEGIES

You can **Use or Make a Picture or Diagram** to help you solve this kind of problem. Look for information in the graph. Match the information with the clues about each student.

SOLVE IT

Look at the graph.

1. To find out how much time Danielle spends on a computer, what do you have to find out first?

2. How can you find the median for sixth grade?

   Is there a number in the middle?

   What is the median for sixth grade?

3. How long does Danielle spend on a computer?

4. To find out how much time Nadia spends, what do you have to find out first?

   What is the median for seventh grade?

5. How much time does Nadia spend?

6. To find out how much time Taylor spends, what do you have to find out first?

   How do you find the mode?

   What is the mode for sixth grade?

7. How much time does Taylor spend?

8. Use the clue for Katie and find out how much time she spends.

9. How much time do Danielle, Nadia, Taylor, and Katie each spend using a computer during the week?

LOOK BACK
- Read the problem again.
- Check over your work.
- Did you answer the question that was asked?
- Does your answer make sense?

Problem Solver II 69

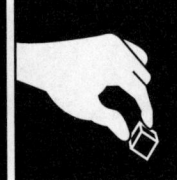

Use or Make a Table
Act Out or Use Objects

Number Cube, Spinner

35 Reed and Cedric are playing a game. On each turn, they toss a number cube and spin on this spinner face. The cube has the numbers 1–6. If the number tossed and the number spun are both even, Reed gets 1 point. If one number is odd and one is even, Cedric gets 1 point. If both numbers are odd, Reed and Cedric both get 1 point. The first player to get 20 points is the winner.

Is this a fair game? Why or why not? You may use the expected probability to explain your answer; or you may play the game and use your results to explain your thinking.

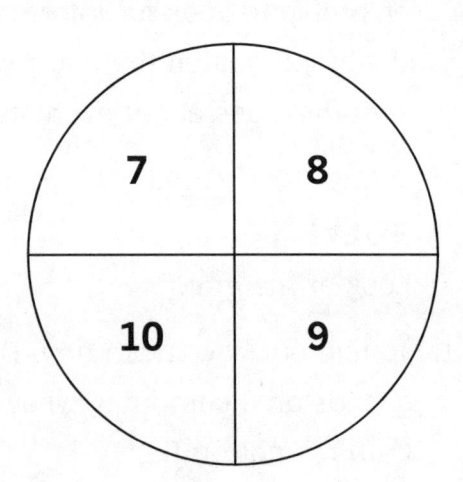

Spinner shows four sections: 7, 8, 10, 9

FIND OUT

What do you have to find out to solve the problem?

What does the problem tell you the players do on each turn?

What do you know about how the players get points and win the game?

CHOOSE STRATEGIES

You can **Act Out or Use Objects** and **Use or Make a Table** to help you solve this kind of problem. Use one number cube and a transparent spinner. Make a table to find all of the possible combinations of even and odd numbers.

70   Problem Solver II

SOLVE IT

Look at the table that has been started.

Possible Outcomes
(EE = even-even; OO = odd-odd;
EO = even-odd)

Number on Cube

	1	2	3	4	5	6
7	OO	EO				

Number on Spinner

1. What numbers are on the cube?

   Does each number have an equal chance of being tossed?

2. What numbers are on the spinner face?

   Does each number have an equal chance of being spun?

3. To find out if the game is fair, you can begin by finding the expected probability of getting each outcome. Fill in the table to find all possible outcomes.

4. How many outcomes are even-even?

   How many are odd-odd?

   How many are even-odd?

5. How many outcomes are possible in all?

6. What is the probability of getting both even numbers?

   What is the probability of getting both odd numbers?

   What is the probability of getting an even number and an odd number?

7. Is the game fair? Why or why not?

LOOK BACK
- Read the problem again.
- Check over your work.
- Did you answer the question that was asked?
- Does your answer make sense?

Problem Solver II

A	B
3	1
7	5

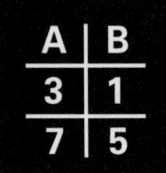

Use or Make a Table
Act Out or Use Objects

Cubes, Spinner

36 Salma and Kenzie are playing a game. They put 1 red cube, 2 blue cubes, and 3 green cubes in a paper bag. On each turn they draw one cube out of the bag and spin the spinner on this spinner face. If the color of the cube matches the color spun, Salma gets 1 point. If the cube does not match the color spun, Kenzie gets 1 point. The first player to have 20 points is the winner.

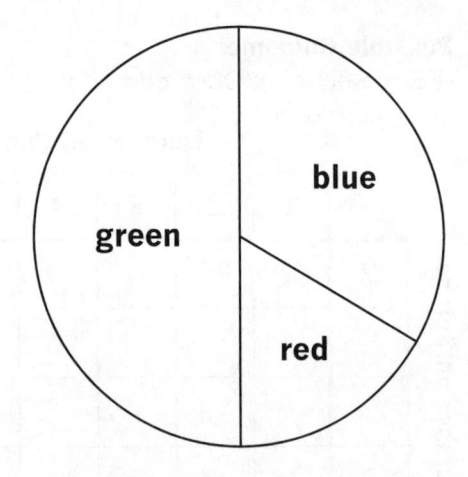

Is this game fair? Why or why not? You may use the expected probability to explain your thinking; or you may play the game and use your results to explain your thinking.

FIND OUT

What do you have to find out to solve the problem?

What does the problem tell you the players do before the game and on each turn?

What do you know about how the players get points and win the game?

CHOOSE STRATEGIES

You can **Act Out or Use Objects** and **Use or Make a Table** to help you solve this kind of problem. Use a bag of colored cubes and a transparent spinner. Make a table to find all of the possible combinations of color drawn and color spun.

SOLVE IT

Look at the table that has been started.

Possible Outcomes
(M = matching; NM = not matching)

Color on Spinner

	green	green	green	blue	blue	red
green	M	M	M	NM		
green						

Color of Cube Drawn

1. What colors are the cubes in the bag, and how many cubes are there of each color?

   Does each cube have an equal chance of being drawn?

   Does each color have an equal chance of being drawn?

   What are the chances of drawing each color?

2. What colors are on the spinner face?

   Does each color have an equal chance of being spun?

   What are the chances of spinning each color, and why?

3. To find out if the game is fair, you can begin by finding the expected probability of getting each outcome. Fill in the table to find all possible outcomes.

4. How many outcomes are matching colors?

   How many outcomes are not matching colors?

5. How many outcomes are possible in all?

6. What is the probability of getting matching colors?

   What is the probability of getting colors that are not matching?

7. Is the game fair? Why or why not?

LOOK BACK
- Read the problem again.
- Check over your work.
- Did you answer the question that was asked?
- Does your answer make sense?

Problem Solver II 73

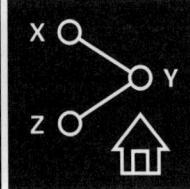

Use Logical Reasoning
Use or Make a Picture or Diagram

37

April, Cindy, Dale, Michael, and Trent are cousins. Each one has a different last name: Byrd, Chang, Clark, Hughes, or Zinn. Each was born in a different year: 1986, 1989, 1990, 1993, or 1999.

- The last two digits of the year in which Trent was born form a prime number. His last name and the last name of another boy cousin begin with the same letter.
- April is 4 years younger than Dale and 9 years older than Cindy. Her last name does not begin with Z.
- Dale is the oldest cousin, and his last name does not begin with C.
- The cousin born in 1993 has the last name Chang.
- The youngest cousin's last name begins with B.

What is each cousin's last name, and when was each one born?

FIND OUT

What do you have to find out to solve the problem?

What are the cousins' names?

What do you know about when they were born?

What do the clues tell you?

CHOOSE STRATEGIES

You can **Use Logical Reasoning** and **Use or Make a Picture or Diagram** to help you solve this kind of problem.

SOLVE IT

Look at the logic chart that has been started.

	Byrd	Chang	Clark	Hughes	Zinn	1986	1989	1990	1993	1999
April										
Cindy										
Dale										
Michael										
Trent										

1. What will you keep track of in the rows?

   In the columns?

2. Begin with the first clue. The last two digits of which year are a prime number?

   Write **yes** in Trent's row to show when he was born. Were any other cousins born in this year?

   Then write **no** in the other spaces in that column. You can also write **no** under the other years in Trent's row.

3. What else does the first clue tell you?

   What names begin with the same letter?

   Is there anyone who cannot have these last names? Write **no** in their rows.

4. Read the rest of the clues. Write **yes** and **no** wherever they belong.

5. What is each cousin's last name, and when was each one born?

LOOK BACK
- Read the problem again.
- Check over your work.
- Did you answer the question that was asked?
- Does your answer make sense?

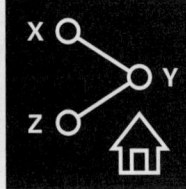

Use Logical Reasoning
Use or Make a Picture or Diagram

38 Bryce, Carlotta, Eva, Jim, and Sara went to the movies on Saturday. Each friend went to a different movie that started at a different time: 12:30, 1:45, 3:30, 4:15, or 5:45. The movies they saw were *Finding Bigfoot 3, Purple People, Midnight Mystery, Lucky Simon,* and *Spy Robots.*

- Jim's movie started 2 hours and 15 minutes before Sara's movie.
- Carlotta's movie started $2\frac{1}{2}$ hours after the movie about Simon.
- The earliest movie is the third in a series.
- *Purple People* started right after *Lucky Simon* and right before *Midnight Mysteries.*
- Eva went to the movie that started right before Jim's movie.

Which movie did each person see, and when did it start?

FIND OUT
What do you have to find out to solve the problem?

What does the problem tell you about who went to the movies and when?

What were the movies they saw?

What do the clues tell you?

CHOOSE STRATEGIES
You can **Use Logical Reasoning** and **Use or Make a Picture or Diagram** to help you solve this kind of problem.

SOLVE IT

Look at the logic chart that has been started.

	Finding Bigfoot 3	Purple People	Midnight Mystery	Lucky Simon	Spy Robots	12:30	1:45	3:30	4:15	5:45
Bryce										
Carlotta										
Eva										
Jim										
Sara										
12:30										
1:45										
3:30										
4:15										
5:45										

1. Think about what you will keep track of in the rows and columns.

2. Begin with the first clue. Which two movie times are 2 hours and 15 minutes apart?

   Where can you write **yes**?

   Write **no** where it belongs in both rows.

3. Read the next clue. Can you find two movie times that are $2\frac{1}{2}$ hours apart?

   Where can you write **yes** for these two movies?

4. Keep reading clues and filling in the chart.

5. Which movie did each person see, and when did it start?

LOOK BACK
- Read the problem again.
- Check over your work.
- Did you answer the question that was asked?
- Does your answer make sense?

Problem Solver II 77

Thinking Questions
Questions to think about as you are solving problems

FIND OUT
What is happening in the problem?
What do I have to find out to solve the problem?
Are there any words or ideas I don't understand?
What information can I use?
Am I missing any information that I need?

CHOOSE STRATEGIES
Have I solved a problem like this before?
What strategies helped me solve it?
Can I use the same strategies for this problem?

SOLVE IT
What information should I start with?
Do I need to add, subtract, multiply, or divide?
How can I organize the information that I use or find?
Is the strategy I chose helpful?
Would another strategy be better?
Do I need to use more than one strategy?
Is my work easy to read and understand? Is it complete?

LOOK BACK
Did I answer the question that was asked in the problem?
Is more than one answer possible?
Is my math correct?
Does my answer make sense? Is it reasonable?
Can I explain why I think my answer is correct?